STUDEBAKER
1933 THROUGH 1942
PHOTO ARCHIVE

STUDEBAKER
1933 THROUGH 1942
PHOTO ARCHIVE

Edited with introduction by
Howard L. Applegate

Iconografix
Photo Archive Series

Iconografix
PO Box 18433
Minneapolis, Minnesota 55418 USA

Library of Congress Card Number 94-79319

ISBN 1-882256-24-7

95 96 97 98 99 00 5 4 3 2 1

Cover and book design by Lou Gordon, Osceola, Wisconsin

Printed in the United States of America

Book trade distribution by Voyageur Press, Inc. (800) 888-9653

PREFACE

The histories of machines and mechanical gadgets are contained in the books, journals, correspondence and personal papers stored in libraries and archives throughout the world. Written in tens of languages, covering thousands of subjects, the stories are recorded in millions of words.

Words are powerful. Yet, the impact of a single image, a photograph or an illustration, often relates more than dozens of pages of text. Fortunately, many of the libraries and archives that house the words also preserve the images.

In the *Photo Archive Series*, Iconografix reproduces photographs and illustrations selected from public and private collections. The images are chosen to tell a story—to capture the character of their subject. Reproduced as found, they are accompanied by the captions made available by the archive.

The Iconografix *Photo Archive Series* is dedicated to young and old alike, the enthusiast, the collector and anyone who, like us, is fascinated by "things" mechanical.

1938 President convertible and club sedans and Pennsylvania Railroad's Broadway Limited, all designed by Raymond Loewy.

INTRODUCTION

The history of Studebaker began in 1852 when two brothers, Henry and Clem opened a South Bend, Indiana blacksmith shop where they specialized in the manufacture of horse drawn farm wagons. In a few years, they were joined by three other brothers, John Mohier, Jacob, and Peter. Soon the family firm evolved into the Studebaker Brothers Manufacturing Company. The company's publicity claimed that the brothers owned and operated the world's largest builder of horse drawn vehicles. The brothers were slow to recognize the fact that the motor vehicle permanently would replace all horse drawn equipment, so they entered the automotive market cautiously and late. Starting in 1902, they made and sold small electric cars, beginning in 1904 they sold cars essentially made to their specifications by the Garford company of Ohio. The brothers established the Studebaker Corporation in 1910 after they acquired the E-M-F Company of Detroit. Thereafter, the Studebakers themselves manufactured all their vehicular products.

The 1920's were the Golden Age of the Studebaker Corporation. Not only was it recognized as a premiere manufacturer of quality cars and trucks, but the company made money and lots of it during the early and middle Twenties. Studebaker tried a number of ventures in the late 1920's and early 1930's to develop a larger portfolio of vehicles that would appeal to a wider customer base. In 1927, the company introduced a new small car, the Erskine, named for Albert R. Erskine, corporate president since 1915.

Officials failed to recognize that the Erskine had two basic flaws: a low torque high axle ratio engine that led to early engine breakdown and a price slightly under one thousand dollars that made the car non-competitive with Ford. consequently, the Erskine died after the 1930 model year. In 1928, Studebaker acquired the Pierce Arrow Motor Car Company to fill the niche above the existing President line with one of the world's greatest luxury cars. President Erskine also invested heavily with Studebaker corporate funds into the White Motor Truck Company to provide heavy duty trucks to complement Studebaker's light and medium commercial lines. Both Pierce and White were sold off cheaply in 1934 fire sales to raise cash for the parent company.

Albert Erskine was undaunted by the failure of the little car bearing his name. In 1932, during the depths of the Great Depression, he introduced another economy car, the Rockne, named for Notre Dame's football coach, who also provided celebrity endorsements for Studebaker products. Unfortunately, Rockne died in an air crash just prior to the car's introduction. Thus, he never learned that the Rockne proved not to be the success that President Erskine expected.

1933 proved to be a difficult year for Studebaker. President Erskine's disastrous policy of distributing millions of dollars in unearned cash dividends to stockholders while the company fell further and further into debt led the company to petition the courts for receivership while at the same time, the assembly lines were stopped. On March 18, 1933, the court appointed Studebaker marketing vice president Paul G. Hoffman, production vice president Harold S. Vance and White Motor chairman Aston Bean as receivers. Within weeks, production of cars and trucks resumed. Erskine remained the nominal president, but he was both clinically depressed and personally insolvent. Consequently a Hoffman and Vance ran the company behind the scenes until July 1, 1933 when Erskine tragically killed himself. The new car styling for 1933 emphasized a slanted grille, bold sweeping skirted fenders, and a new rear end treatment. Hoffman and Vance added a fourth series, the President Speedway, the most expensive cars, which lasted only one model year. Sales for the 1933 year were slightly lower than 1932.

Hoffman and Vance showed their management skills in 1934 by securing a seven million dollar line of credit and opening a second manufacturing facility in Los Angeles, California. Hoffman used some of the bank funding in a multi-million dollar marketing blitz that had as its theme, "From the Speedway comes the Stamina." Vance's contributions included the hiring of styling consultant Raymond Loewy who developed a special sedan, the Land Cruiser, based on the Studebaker sponsored 1933 Pierce prototype Silver Arrow, with its rear fender skirts and four piece rear window treatment and the introduction of mid-year design changes with the "Year Ahead Models" that featured horizontal hood louvers instead of the typical vertical treatment. Loewy basically modified existing designs until the 1937 models. Car production rose slightly to over 50,000 units.

The 1935 cars emphasized new technology rather than styling where the old streamlined effect now had a more chiseled look. Vance's engineers produced many new innovations including the Planar independent front wheel suspension that the marketing department called "Miracle Ride," vacuum powered hydraulic brakes, overdrive on the Presidents and the "Hill Holder" that locked the brakes when the driver depressed the clutch. In spite of good design and modern technology, sales did not increase.

The depression slowed somewhat in 1936 and consequently Studebaker sales increased. The cars had a new body design with interiors appointed by Helen Dryden, a famous interior decorator. The bodies featured all steel construction with turret type tops. The engineers provided an independent sprung front wheel system marketed as "Startix". The Commander series was dropped from the Studebaker lineup. The 1937 Loewy cars had a more modern design with a pointed grille and Studebaker's first alligator hood opening. The company tried to enhance their sales of commercial vehicles by creating the Coupe Express that combined a passenger car front end with a pickup bed and tailgate. This beautiful vehicle never sold well and was dropped after 1939. More cars were sold in 1937 than in 1936. Loewy continued to work on streamlining car styling and the 1938's were much bolder than the 1937's. Studebaker officials recognized the European facts of life in the development of dictatorships in Germany, Italy, and Russia by dropping the Dictator series. The replacement originally was labeled the Studebaker Six, but it evolved into a new Commander Six series starting on December 3, 1937. The engineers provided a vacuum accentuated optional "Miracle Shift" that made shifting much smoother. A convertible sedan, essentially a one year body style, was offered in both the Commander and President series. The depression unexpectedly got worse in 1938 and sales declined.

1939 is a major year in the history of the Studebaker Corporation. Not only were the presidents and Commanders beautifully redesigned, but Hoffman and Vance restored the company to financial health with their mid-year introduction of the 1939 Champion series, a high quality small car. The Champion only cost about twenty five dollars more than the "Big Three" low priced cars plus it had pleasing styling, good fuel economy and terrific performance. Hoffman and Vance were the ultimate risk takers given Studebaker's previous small car fiascoes with the Erskine and the Rockne and given their investment of nearly five million dollars. The results for 1939 showed them to be consummate risk takers and not wild gamblers as nearly 73,000 of the Champions were sold in the first year. Given the company's interest in commercial vehicles, it should not be surprising that Champions could be made into sedan deliveries with the purchase of a special kit from the factory or into pickups by having a coupe converted by Edwards.

1940 was another banner year for Studebaker. Hoffman invested much money into Champion advertising while the Studebaker exhibit at the New York World's Fair showed many tourists why they should buy Champions. Sales of new cars especially Champions increased. Consequently, the Champion line was a major money maker in 1940. To attract the public to the standard sized cars, the styling was streamlined and Loewy introduced a very attractive two tone effect that Studebaker called "De Lux Tone". The convertible sedan was no longer a stock model, but elite customers who wanted such a car could have a sedan converted by the Derham Coach Company.

The 1941 cars were streamlined again by Loewy who produced what some Studebaker historians believe was the best styling of the Thirties and early Forties. The redesigned Champions as executed by Loewy appeared to be much larger than their predecessors with the result that 1941 Champion production was the largest in corporate history. The standard sized cars had hidden running boards in a smooth treatment of the sheet metal. At mid-year, Studebaker introduced a new body style called the Sedan-Coupe, a really stylish improvement of the old two door sedan, which featured a thoroughly modern one piece windshield. Sales for 1941 were over the one hundred million dollar mark for the first time since 1929. For the 1942 models, Studebaker engineers offered a "Turbo-matic" drive optional on Presidents and Commanders that eliminated the clutch and reduced gear shifted to an absolute minimum similar in nature to Chrysler's Fluid Drive. World War II interrupted passenger car production in January of 1942 after only 9,285 had been built.

The war had a tremendous impact on Studebaker. During the period 1942-1945, Studebaker was a primary national defense contractor and produced nearly 200,000 army trucks, aircraft engines and numerous Weasels, which the army called an amphibious troop carrier. On its defense cost plus contracts, Studebaker earned over one billion dollars, prompting the company to pay its first cash dividend to stockholders since 1933.

The percentage of profits were modest, however, and Studebaker did not become a major war profiteer. Based on the company's sales records for the 1939-1941 era, some Studebaker historians believe that had war not come, the company would have done well in the 1942-1945 period selling civilian vehicles. The bottom line result, however, was that Studebaker had a reserve fund with which to enter the postwar automotive arena in 1946.

1933 Six Suburban with body by Cantrell.

1933 Commander two-passenger coupe.

1933 Commander four-door sedan.

1933 Commander five-passenger sedan.

1933 Commander Eight Regal six-passenger sedan.

1933 Commander Regal St. Regis Brougham for five-passengers.

1933 President Speedway four-passenger coupe.

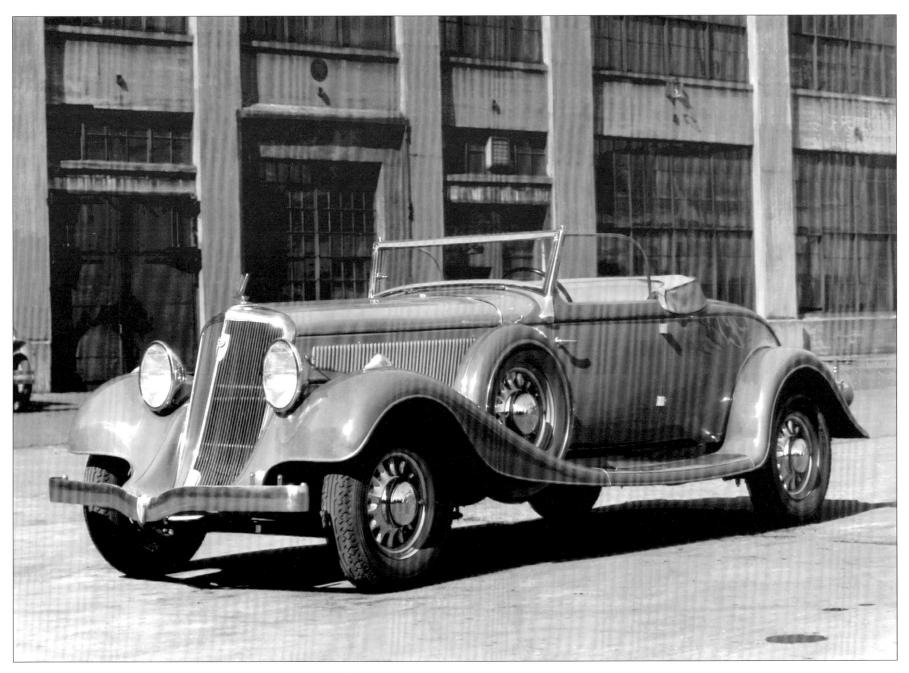

1933 President 8 Regal four-passenger roadster.

1933 President Regal St. Regis Brougham.

1933 President Berline.

1933 President Speedway seven-passenger limousine.

1933 President Speedway State convertible sedan for seven-passengers.

1933 Commander Six convertible sedan.

Assortment of new 1934 Studebakers on a typical car carrier.

1934 Dictator Regal six-passenger sedan and Kate Smith at Sioux City, Iowa.

1934 Dictator St. Regis Brougham mini-ambulance.

Loading 1934 Studebakers into a New York Central boxcar at South Bend.

1934 Dictator Suburban with body by U.S.B.&F.

1934 Dictator taxicab at South Bend railroad station.

1934 Dictator funeral service car with body by Superior.

1934 Commander St. Regis Brougham.

1934 Commander four-door sedan in Kansas City park.

1934 President Regal two-passenger coupe.

1934 President cruising sedan entered in Union Oil Company endurance run.

1934 President Regal Berline limousine in Des Moines, Iowa.

1934 Land Cruiser with Loewy special two-tone treatment.

1934 President Land Cruiser, designed by Raymond Loewy.

1934 President Samaritan ambulance with body by Superior.

1934 President Arlington funeral car with body by Superior.

1934 "Year Ahead" Dictator Regal convertible roadster for five-passengers.

1934 "Year Ahead" Commander Regal five-passenger sedan.

1934 "Year Ahead" Commander five-passenger sedan.

1934 "Year Ahead" Commander Regal Land Cruiser in McKeesport, Pennsylvania.

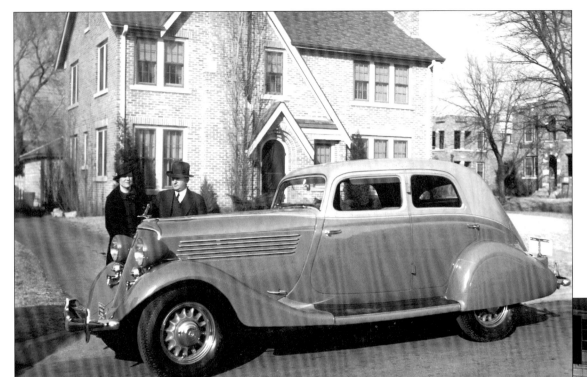

1934 "Year Ahead" President Land Cruiser in Oklahoma City.

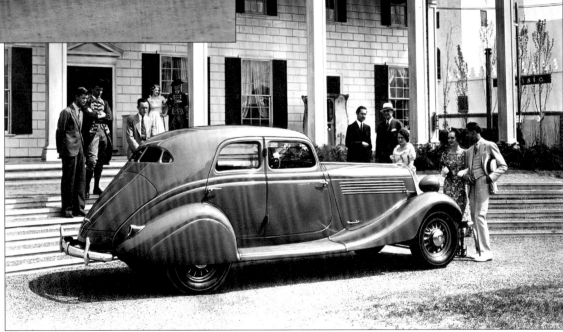

1934 "Year Ahead" President Land Cruiser with single-tone paint.

42

1935 Studebaker exhibit at the New York Auto Show, in the Grand Central Palace.

1935 Dictator three-passenger coupe.

1935 Dictator Custom Cruising sedan and actor Douglas Fairbanks (right) in Shanghai, China.

1935 Commander Land Cruiser and child actor Mickey Rooney.

1935 Dictator funeral service car with body by Superior.

1935 Commander three-passenger coupe at Philadelphia Navy yard.

1935 Commander six-wheeled sedan.

1935 Commander Mission funeral car with body by Superior.

1935 President Custom six-passenger sedan.

1935 President Land Cruiser.

1935 President Land Cruiser at the Toledo Auto Show.

1935 President cruising sedan with bullet proof tire deflectors and bullet proof radiator.

1935 President Samaritan ambulance with body by Superior.

1935 President Elmhurst coroner's service car with body by Superior.

1935 President Westminster funeral car with body by Superior.

56

1936 Studebaker courtesy car used by Marx Brothers at Cleveland Municipal Stadium.

1936 final assembly line.

1936 Dictator two-tone. Demonstrating the strength of the all steel roof.

1936 Dictator three-passenger coupe.

1936 Dictator three-passenger coupe at Wyoming dealership.

1936 Dictator St. Regis sedan.

1936 Dictator Six St. Regis mini-ambulance.

1936 President three-passenger coupe.

1936 President St. Regis sedan.

1936 President cruising sedan.

1936 President five-passenger coupe and singer Lawrence Salerno.

1936 President Westminster funeral car with body by Superior.

1936 President combination coach professional car with body by Superior.

1936 President ambulance with body by Superior.

1937 Dictator St. Regis six-passenger sedan.

1937 Dictator cruising sedan.

1937 Dictator Arlington funeral car with body by Superior.

1937 Dictator Suburban with body by U.S.B.&F.

1937 Dictator taxicabs outside the Philadelphia Inquirer building.

1937 Coupe Express.

1937 Coupe Express with wooden stake body.

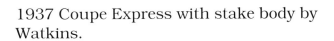

1937 Coupe Express with stake body by Watkins.

1937 prototype sedan delivery with 1936 wheels and a dummy grille. Apparently, none were built.

1937 President five-passenger coupe.

1937 President cruising sedan with optional sun roof.

1937 President cruising sedan and trans-Atlantic flyer Harry Richman, who also served as a civilian fire chief.

1937 President cruising sedan with German racer Bernd Rosemeyer.

1937 President cruising sedan at Chicago airport.

1937 President Westminster funeral car with body by Superior.

1938 Commander State six-passenger club sedan.

1938 Commander Six four-door sedan.

1938 Commander Six Suburban with body by U.S.B.&F.

1938 Commander convertible sedan.

1938 Commander cruising sedan with driver education equipment.

1938 Commander Six Coupe Express.

A 1937 and 1938 Coupe Express with canvas bodies used by the Canadian Royal Mail Service.

1938 President State convertible sedan.

1938 President State six-passenger
club sedan.

1938 President Arlington funeral car with body by Bender.

Hoffman and Vance with 1939 Champion Custom cruising sedan.

1939 Champions at a Jacksonville, Florida dealership.

1939 Champions at the Canadian National Exhibition, Toronto, Canada.

1939 Champion Custom cruising sedan.

1939 Champion pickup coupe with modification by Edwards Iron Works of South Bend.

1939 Commander Coupe Express at Youngstown, Ohio dealership.

1939 Commander Coupe Express.

1939 Commander Coupe Express used by Texas dealer.

1939 Commander station wagon with body by McAvoy and Sons.

1939 Commander convertible sedan prototype.

1939 Commander Custom.

1939 Commander cruising sedan.

1939 President Club Sedan.

Armored 1939 President cruising sedan, with bullet deflectors on front wheels, for export to China.

1940 Champion De Luxe club sedan with Wilbur Shaw on right. *Photograph used with permission of the Indianapolis Motor Speedway.*

1940 Champion De Luxe three-passenger coupe and actress Judy Garland.

1940 Champion five-passenger coupe.

1940 Champion De Luxe club sedan.

1940 Champion police car in Glace Bay, Canada.

1940 Champion Custom cruising sedan for five-passengers.

1940 Champion delivery sedan.

1940 Champion sedan delivery. Demonstration of the cubic and load capacities.

1940 Champion sedan deliverys outside a dealership.

1940 Commander three-passenger coupe.

1940 President six-passenger club sedan.

1940 President cruising sedan.

1941 Champion Custom De Luxe three-passenger coupe in Los Angeles, California.

120

1941 Champion six-passenger cruising sedan.

1941 Champion Coupe Express.

1941 Commander De Luxe-Tone six-passenger cruising sedan.

1941 Commander Skyway six-passenger sedan coupe.

1941 Commander De Luxe-Tone Land Cruiser.

1941 President Skyway sedan coupe.

1941 President Skyway sedan coupe.

1941 President Skyway sedan coupe.

1941 President De Luxe-Tone Land Cruiser six-passenger sedan.

1941 President De Luxe-Tone six-passenger cruising sedan at Notre Dame stadium.

1941 President Skyway Land Cruiser.

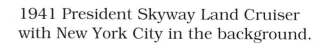

1941 President Skyway Land Cruiser
with New York City in the background.

1941 President convertible sedan, custom built by Derham, with a 1940 front end and a 1941 rear deck.

Fleet of 1942 Champion coupes employed by the New York police.

1942 Champion club sedan for five-passengers.

1942 Champion Suburban, probably one of a kind.

1942 Commander Custom cruising sedan.

Bob Hope and Jerry Colona with a 1942 Commander Skyway sedan coupe.

1942 Sedan coupe power demonstration with Milwaukee Road locomotive.

1942 President Skyway Land Cruiser.

1942 President Skyway sedan coupe.

1942 President Skyway sedan. The last car to be manufactured before military production.

The photographs reproduced in this book are from the collection of Shelby and Howard Applegate. Many of the photographs, as well as thousands of others of American passenger cars and commercial vehicles, are available at wholesale or retail. Studebaker car and truck original sales literature, owner manuals, tune-up charts, paint charts, and stock certificates also are available, as is similar material on most popular American and foreign marques. Inquiries from buyers and sellers of automobilia are invited.

APPLEGATE AND APPLEGATE
PO BOX 260
ANNVILLE, PENNSYLVANIA 17003

EVENING TELEPHONE: (717) 964-2350

The Iconografix Photo Archive Series includes:

JOHN DEERE MODEL D Photo Archive	ISBN 1-882256-00-X
JOHN DEERE MODEL A Photo Archive	ISBN 1-882256-12-3
JOHN DEERE MODEL B Photo Archive	ISBN 1-882256-01-8
JOHN DEERE 30 SERIES Photo Archive	ISBN 1-882256-13-1
FARMALL REGULAR Photo Archive	ISBN 1-882256-14-X
FARMALL F-SERIES Photo Archive	ISBN 1-882256-02-6
FARMALL MODEL H Photo Archive	ISBN 1-882256-03-4
FARMALL MODEL M Photo Archive	ISBN 1-882256-15-8
CATERPILLAR THIRTY Photo Archive	ISBN 1-882256-04-2
CATERPILLAR SIXTY Photo Archive	ISBN 1-882256-05-0
CATERPILLAR MILITARY TRACTORS VOLUME 1 Photo Archive	ISBN 1-882256-16-6
CATERPILLAR MILITARY TRACTORS VOLUME 2 Photo Archive	ISBN 1-882256-17-4
TWIN CITY TRACTOR Photo Archive	ISBN 1-882256-06-9
MINNEAPOLIS-MOLINE U-SERIES Photo Archive	ISBN 1-882256-07-7
HART-PARR Photo Archive	ISBN 1-882256-08-5
OLIVER TRACTORS Photo Archive	ISBN 1-882256-09-3
HOLT TRACTORS Photo Archive	ISBN 1-882256-10-7
RUSSELL GRADERS Photo Archive	ISBN 1-882256-11-5
MACK MODEL AB Photo Archive	ISBN 1-882256-18-2
MACK MODEL B 1953-66 Photo Archive	ISBN 1-882256-19-0
MACK FC, FCSW & NW 1936-1947 Photo Archive	ISBN 1-882256-28-X
MACK EB, EC, ED, EE, EF, EG & DE 1936-1951 Photo Archive	ISBN 1-882256-29-8

LE MANS 1950: THE BRIGGS CUNNINGHAM CAMPAIGN Photo Archive	ISBN 1-882256-21-2
SEBRING 12-HOUR RACE 1970 Photo Archive	ISBN 1-882256-20-4
IMPERIAL 1955-1963 Photo Archive	ISBN 1-882256-22-0
IMPERIAL 1964-1968 Photo Archive	ISBN 1-882256-23-9
STUDEBAKER 1933-1942 Photo Archive	ISBN 1-882256-24-7
STUDEBAKER 1946-1958 Photo Archive	ISBN 1-882256-25-5
AMERICAN SERVICE STATIONS 1935-1943 Photo Archive	ISBN 1-882256-27-1
CASE TRACTORS 1912-1959 Photo Archive	ISBN 1-882256-32-8
FORDSON 1917-1928 Photo Archive	ISBN 1-882256-33-6

Available Late 1995

MACK MODEL B 1953-1966 VOL.2 Photo Archive	ISBN 1-882256-34-4
MACK FG-FH-FJ-FK-FN-FP-FT-FW 1937-1950 Photo Archive	ISBN 1-882256-35-2
DODGE TRUCKS 1929-1947 Photo Archive	ISBN 1-882256-36-0
DODGE TRUCKS 1948-1961 Photo Archive	ISBN 1-882256-37-9
MACK EH-EJ-EM-EQ-ER-ES 1936-1950 Photo Archive	ISBN 1-882256-39-5
MACK LF-LH-LJ-LM-LT 1940-1956 Photo Archive	ISBN 1-882256-38-7
STUDEBAKER TRUCKS 1928-1940 Photo Archive	ISBN 1-882256-40-9
STUDEBAKER TRUCKS 1941-1964 Photo Archive	ISBN 1-882256-41-7

The Iconografix Photo Archive Series is available from direct mail specialty book dealers and bookstores throughout the world, or can be ordered from the publisher.

For information write to:
Iconografix
PO Box 609
Osceola, Wisconsin 54020 USA

Telephone: (715) 294-2792
(800) 289-3504 (USA and Canada)
Fax: (715) 294-3414